Grade 5 · Unit 3

Inspire Science

Earth's Interactive Systems

McGraw Hill Education

Mheducation.com/prek-12

STEM McGraw-Hill is committed to providing
instructional materials in Science, Technology,
Engineering, and Mathematics (STEM) that give all
students a solid foundation, one that prepares them
for college and careers in the 21st century.

Send all inquiries to:
McGraw-Hill Education
8787 Orion Place
Columbus, OH 43240

ISBN: 978-0-07-699677-3
MHID: 0-07-699677-8

Printed in the United States of America.

7 8 9 10 11 12 LWI 26 25 24 23 22 21

Table of Contents
Unit 3: Earth's Interactive Systems

Earth's Other Systems

Earth's Water System

ENCOUNTER
THE PHENOMENON

How does water collect in ecosystems?

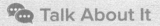

GO ONLINE
Check out *Collecting Water* to see the phenomenon in action.

💬 Talk About It

Look at the photo and watch the video of water collecting on the insect. What questions do you have about the phenomenon? Talk about them with a partner.

Did You Know?

Some plants and animals use their parts to pull water vapor from the air around them.

Design a Rainwater Collection System

How can rainwater be collected to be reused? You have been hired as a hydraulic engineer to design a rainwater collection system in your area. A hydraulic engineer works with a variety of people to design structures that can store water. At the end of this module, you will build and test your model to communicate the advantages and disadvantages of using a rainwater collection system.

Lesson 1
Water Distribution on Earth

Lesson 2
Human Impact on Water Resources

Lesson 3
Effects of the Hydrosphere

Hydraulic engineers need to make sure the storage structures they design do not interfere with the surrounding land and that the water that is collected is safe to use.

STEM Module Project

Plan and Complete the Engineering Challenge You will use what you learn to design a rainwater collection system.

What Covers Earth?

Three friends wondered what covered most of Earth's surface.
They each had different ideas. This is what they said:

Mia: *I think Earth is covered mostly by ice and snow.*

Sam: *I think Earth is covered mostly by water.*

Cate: *I think Earth is covered mostly by land.*

Who do you agree with the most? _____

Explain why you agree.

_____

_____

_____

_____

_____

_____

You will revisit the Page Keeley Science Probe later in the lesson.

Water Distribution on Earth

ENCOUNTER
THE PHENOMENON

Where is water located in this image?

🔘 GO ONLINE

Check out *Water on the Mountains* to see the phenomenon in action.

Look at the photo and watch the video of the Sierra Nevada mountains. What questions do you have about the phenomenon? Record or illustrate your thoughts below.

Did You Know?

In order for a landform to be classified as a mountain in the United States, it has to be about 300 meters (just under 1,000 feet) taller than the land surrounding it.

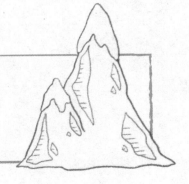

Data Analysis

Usable Water

Think about where the water appears on the mountains. Water covers seventy-percent of Earth's surface. All living things need water in the form of fresh water, but water also appears as salt water on Earth.

Make a Prediction How much of the total water on Earth is fresh water compared to salt water?

_____

_____

_____

_____

Carry Out an Investigation

1. **MATH Connection** Look at the diagram on the next page.

2. What type of water is most of the water on Earth?

_____

_____

_____

_____

3. According to the diagram, what types of water sources are sources of fresh water?

_____

_____

_____

_____

_____

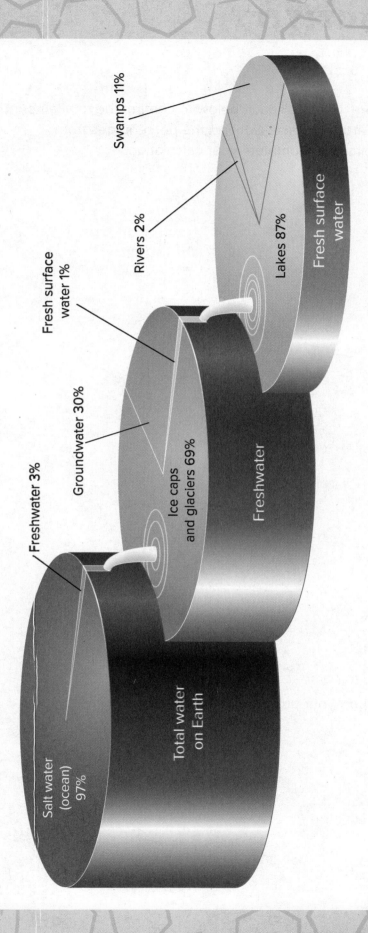

Swamps 11%

Rivers 2%

Lakes 87%

Fresh surface water

Fresh surface water 1%

Groundwater 30%

Ice caps and glaciers 69%

Freshwater

Freshwater 3%

Salt water (ocean) 97%

Total water on Earth

INQUIRY ACTIVITY

Communicate Information

4. **MATH Connection** Use the space below to confirm the total amount of fresh water on Earth's surface. Convert the percentages from the diagram to decimals to complete your calculation.

5. Do the results support your prediction? Explain why or why not.

_____

_____

_____

_____

_____

MAKE YOUR CLAIM

How much of Earth's surface water is available to living things?

Make your claim. Use your investigation.

CLAIM

_____ is **available** to living things.

Cite evidence from the activity.

EVIDENCE

The investigation showed that _____.

Discuss your reasoning as a class. Tell about your discussion.

REASONING

The evidence supports the claim because _____.

You will revisit your claim to add more evidence later in this lesson.

Look for these
words as you read:

glacier

groundwater

ice caps

reservoir

storage

Water on Earth

The water found on Earth makes up the hydrosphere. About 97 percent of Earth's surface water is salt water found in oceans. We cannot drink salt water or use it to grow crops. For those activities, we need fresh water.

Only about 3 percent of Earth's water is fresh water. Most of this fresh water is frozen in the form of permanent snow cover, glaciers, and ice caps. A **glacier** is a thick sheet of ice. A giant **ice cap** covers Antarctica—the continent at the South Pole. This frozen water accounts for about 69 percent of Earth's fresh water. Another 30 percent is groundwater. **Groundwater** is water stored in the cracks and spaces between particles of soil and underground rocks. Less than 1 percent is running water, such as rivers, and standing water, such as lakes. A tiny bit of Earth's water is found in the atmosphere as water vapor.

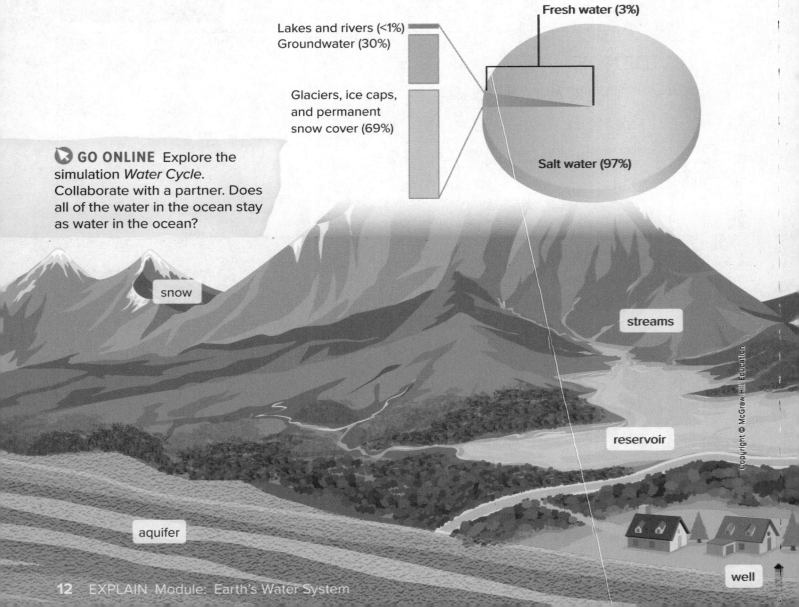

Lakes and rivers (<1%)
Groundwater (30%)

Fresh water (3%)

Glaciers, ice caps, and permanent snow cover (69%)

Salt water (97%)

GO ONLINE Explore the simulation *Water Cycle*. Collaborate with a partner. Does all of the water in the ocean stay as water in the ocean?

snow

streams

reservoir

aquifer

well

Freshwater Sources

GO ONLINE Watch the video *Groundwater* to learn more about how it is stored and used.

There are three main sources of usable fresh water.

Groundwater When water seeps into soil, it enters groundwater aquifers, or underground layers of rock or soil that water can pass through. As water flows through an aquifer, it eventually reaches a layer of rock that it cannot move through. Fresh water builds up on top of this rock. It can be reached by drilling or digging into the ground and pumping the water up through a well.

Running Water Many cities and towns are built next to sources of running water, such as streams or rivers. Thousands of fresh water rivers cross Earth's surface. Running water provides a source of fresh water for homes, farms, and businesses.

Standing Water Bodies of standing fresh water, such as lakes and reservoirs, are also sources of usable fresh water. A **reservoir** is an artificial lake built for storage of water. Reservoirs are usually made by building a dam on a river. Water is stored behind the dam and released when needed. **Storage** is the process of water being stored on Earth's surface, in the ground, or as a water feature.

REVISIT Revisit the Page Keeley Science Probe on page 5.

PAGE KEELEY
**SCIENCE
PROBES**

dam

river

INQUIRY ACTIVITY

Data Analysis

Where Water Is Found

Earth is covered with water. Earth's water features are found in many shapes and sizes.

State the Claim How can we make a model to compare the amounts of salt water and fresh water on Earth?

Materials

colored pencils

graph paper

_____

_____

_____

Carry Out an Investigation

1. Use information from the lesson, such as the diagram from the *Usable Water* activity, as well as information you have read and online resources to make a graph that shows the amount of salt water, frozen fresh water, and available fresh water on Earth in the space on the next page.

Communicate Information

2. **MATH Connection** How does your graph represent the availability of fresh water on Earth?

_____

_____

_____

_____

■ COLLECT EVIDENCE

Revisit your claim on page 11. Use what you have learned as evidence for the reasoning that supports your claim.

ENVIRONMENTAL Connection

SEP DCI CCC

How does the amount of usable fresh water available affect life on Earth?

What Does a Sustainability Specialist Do?

Sustainability Specialists find ways to reduce the use of electricity and other resources for businesses. I reduce a company's carbon output by ensuring that their vehicles are electric. Companies save money by following plans that I have created to reduce the amount of water and electricity used. I may also organize volunteer opportunities for a business, such as a company-wide cleanup day. Volunteers work together to reduce the amount of trash in the community where they are located. Another responsibility of mine is to design a company's product line so it uses only recyclable materials in production. A sustainability specialist's main goal is saving resources. We find as many ways possible to go green!

It's Your Turn

How does the work of a sustainability specialist affect Earth's water? How could a sustainability specialist and a hydraulic engineer work together?

Write a poem to describe how the work of
a sustainability specialist directly affects the sources of water on Earth.
Recite your poem to a partner.

Review

EXPLAIN
THE PHENOMENON

| Where is water located in this image?

Summarize It

Use evidence from the lesson to explain how water is distributed on Earth's surface.

_____

_____

_____

_____

_____

_____

_____

_____

_____

REVISIT
PAGE KEELEY SCIENCE PROBES

Revisit the Page Keeley Science Probe on page 5. Has your thinking changed? If so, explain how it has changed.

Three-Dimensional Thinking

1. All water on Earth is recycled through _____.

   A. the water cycle

   B. pockets of nitrogen

   C. exhaled gases

   D. dead plant and animal matter

2. **True or False** The hydrosphere covers about 70% of Earth's surface.

   A. True

   B. False

3. About _____ of the world's water is salty ocean water.

   A. 12 percent

   B. 43 percent

   C. 47 percent

   D. 97 percent

Extend It

Think about how you modeled the amounts of fresh water and salt water on Earth. How could you model how much of the water in your state is salt water and how much is fresh water? Make a plan to research and build your model below.

_____

_____

_____

_____

KEEP PLANNING
STEM Module Project
Engineering Challenge

Now that you have learned about water distribution on Earth, go to your Module Project to consider this information as you design a rainwater collection system.

Human Impact on Water Resources

Humans do things that can affect the environment. Circle the things that are examples of ways humans can affect water resources.

Growing crops for food	Building a garden wall to prevent erosion	Watering flowers at night
Turning the faucet off when not in use	Building a bridge over a river	Designing a rainwater collection system
Paving a gravel driveway	Planting forests	Using pesticides to treat garden plants

Explain your choices.

_____

_____

_____

_____

_____

_____

_____

You will revisit the Page Keeley Science Probe later in the lesson.

Human Impact on Water Resources

ENCOUNTER
THE PHENOMENON

How were humans involved in this photo?

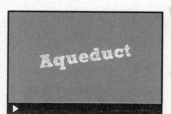

 GO ONLINE

Check out *Aqueduct* to see the phenomenon in action.

Look at the photo and watch the video of the aqueduct. What questions do you have about the phenomenon? Talk about them with a partner. Record or illustrate your thoughts below.

Did You Know?

Nearly 97 percent of Earth's water is found in oceans.

INQUIRY ACTIVITY

Hands On

Test the Waters

Water is essential for all living things on Earth, but not all water on Earth is readily able to be used. Think about what might make water unsafe to use.

Make a Prediction How can we test water to know if it is safe to use?

_____

_____

_____

🔲 **GO ONLINE** Use the *What Is pH?* personal tutor to learn about testing pH levels.

Carry Out an Investigation

BE CAREFUL Wear all required safety gear as directed by your teacher. Do not drink or smell the water samples.

1. Line up the water samples in order from left to right in front of you, each on top of a small piece of paper towel.

2. Dip the pH testing strip halfway into the water. Compare the result on the testing strip to the pH indicator key.

3. **Record Data** Record this information in the data table on the next page.

4. Repeat steps two and three with the rest of your samples. Be sure to use a new pH testing strip for each sample.

Materials

safety goggles

6 small plastic cups with water samples prepared by your teacher

pH testing strips

paper towels

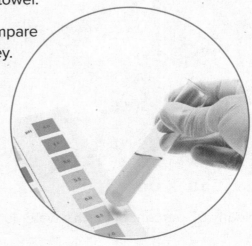

Sample	pH Level	Appearance of Sample
1.		
2.		
3.		
4.		
5.		
6.		

Communicate Information

5. The best pH level for drinking water is 7, but anywhere from 6.5 to 9.5 is safe. Assuming the water samples do not contain anything else that is harmful, how many samples could be safe to drink?

_____

_____

6. As your teacher reveals the contents of each sample, record them in the table. Would you drink the samples that are considered safe? Explain.

_____

_____

_____

7. Do the results support your prediction? Explain.

_____

_____

_____

 Talk About It

Think about the appearance of each of the samples. Is it possible to know if water is safe to drink just by looking at it? Is testing for pH enough to know if water is safe? Talk about it with a partner.

Humans Affect Water

VOCABULARY

Look for these
words as you read:

algal bloom

acid rain

conservation

People affect the environment every day. Sometimes these effects are negative and can harm the environment, such as pollution. Pollution is any harmful substance that affects Earth's resources.

Heavy rains can wash fertilizers used on farms and lawns into lakes, rivers, and streams. These fertilizers can negatively impact water quality and cause toxic kinds of algae to grow. This can result in something called an **algal bloom**, a sometimes harmful increase in the amount of algae found in water. Oil spills are another cause of water pollution. In 2010, the worst oil spill in United States history occurred when an oil rig in the Gulf of Mexico exploded, releasing 4.9 million barrels of oil into the gulf. That amount of oil could fill over 300 Olympic-sized pools!

Even though some human activities harm natural resources, there are many people who work hard to protect them. In 1974, the United States government passed a law to help protect our water. The Safe Drinking Water Act protects drinking water and water resources. Farmers are now using safer, more natural ways of controlling pests and providing nutrients to plants.

Zebra mussels can filter feed. They eat the good algae but release the organisms that contribute to the algal bloom back into the water intact.

The green water is evidence of an algal bloom. Algal blooms can be harmful.

Conservation

Earth does not have an unlimited supply of natural resources. Many resources are being used more quickly than nature can replace them. Humans can help slow the use of natural resources through conservation. **Conservation** is the practice of using resources wisely. The "three Rs" guide people in how to conserve resources. They are reduce, reuse, and recycle.

In terms of conserving water resources, reducing our usage means to use less water. This can mean taking shorter showers and turning the water off while we are brushing our teeth.

Reusing water resources means using it for something else or using it more than once. Rainwater can be collected to be used to water plants.

Recycling water is a complicated process that is done at water treatment facilities. Water that is collected through pipes in homes and offices can be recycled to be used again.

💬 Talk About It

How could you investigate the amount of water you use during daily activities? Talk about it with a partner.

> 📹 **GO ONLINE** Watch the video *Caring for Earth's Water* to learn about more ways to conserve water resources.

Use water-conserving showerheads and take shorter showers.

Do not leave water running when you are not using it.

If you use a dishwasher, use a water-saving model and do not run it unless it is full.

Fix leaking pipes or faucets.

Use a water-saving washing machine and wash full loads of clothes.

Grow plants that do not require frequent watering, and water your plants after dark so the water does not evaporate.

REVISIT  PAGE KEELEY SCIENCE PROBES   Revisit the Page Keeley Science Probe on page 21.

INQUIRY ACTIVITY

Hands On

Effects of Acid Rain

Human activity, such as burning fossil fuels, can lead to acid rain. You will investigate if there is an effect from acid rain on rock material, such as chalk, which is a form of limestone.

Make a Prediction What does acid rain do to some "rocks"?

_____

_____

Carry Out an Investigation

BE CAREFUL Wear safety goggles to protect your eyes.

1. Place a piece of chalk on each of the paper plates. Label one plate *vinegar* and one plate *water*.

2. Using the dropper, carefully place two drops of vinegar solution directly onto one end of the chalk labeled *vinegar*. Use the second dropper to place two drops of water on the chalk labeled *water*.

3. Observe the reaction. Record your observations in the table below. Repeat steps 2 and 3 a total of six times.

	Observations
Chalk with Vinegar Solution	
Chalk with Water	

Materials

safety goggles

2 paper plates

marker

small cup of vinegar solution

small cup of water

2 pieces of chalk

2 droppers

Communicate Information

4. Did your results support your prediction? Explain.

_____

_____

Copyright © McGraw-Hill Education (3)Ken Karp/McGraw-Hill Education, (6 7)Jacques Cornell/McGraw-Hill Education, (others)Ken Cavnanagh/McGraw-Hill Education

Effects of Acid Rain on Ecosystems

◐ GO ONLINE Explore *Effects of Acid Rain* to see more examples of how acid rain affects Earth.

Acid rain results when gases, such as sulfur dioxide and nitrogen oxide, are released into the atmosphere. A small portion of the gases that cause acid rain comes from natural sources such as volcanoes. Most of the gases come from burning fossil fuels. These gases react with water, oxygen, and other chemicals to form acids. These acids mix with water before falling as precipitation.

The effects of acid rain are seen mostly in water environments, such as streams, lakes, and marshes. It can be harmful to fish and other wildlife. If something harms one part of an ecosystem — one species of plant or animal, the soil, or the water — it can have an impact on everything else.

Some types of plants and animals can live in acidic waters. Others are acid-sensitive and will be lost as acid rain enters the ecosystem. Some acidic lakes have no fish. Even if a species of fish or animal can tolerate moderately acidic water, the animals or plants it eats might not.

Acid rain has destroyed this forest.

1. What can be done to reduce or prevent further acid rain damage?

_____

_____

_____

INQUIRY ACTIVITY

Human Impact

Choose one way humans impact water resources and perform research about it. Record your question that you want to answer.

Ask a Question

_____

_____

Define the activity as having a positive or negative effect on the environment. Research community groups that support or are critical of the activity. Record your notes below.

_____

_____

_____

_____

_____

_____

Communicate Information

 How could you plan a campaign or design a poster to **communicate information** to support the positive impact of the activity or reduce the negative impact of the activity that will **affect Earth's water system**?

_____

_____

_____

_____

Organize your information on a poster to present to the class.
Sketch a draft of your poster below.

How Could You Become an Agricultural Inspector?

Agricultural Inspectors have at least a two-year college degree. Their degrees are typically in a science or related field, such as biology or animal science. They take courses in biology, with a heavy focus on plants, animals, and agriculture, which is the science of farming.

After earning a degree, agricultural inspectors receive most of their training when they begin work. They learn about state and national laws and regulations for producing and harvesting food. They are taught how to write reports on what they observe during their visit. They also learn how to collect and analyze samples from plants, animals, soil, and water taken from around the farm.

It's Your Turn

How would a hydraulic engineer and an agricultural inspector work together? Perform research to find out.

Read the Investigator article *Studying Pollution.* What events are contributing to the pollution in oceans and other bodies of water? What choices can humans make to help the environment?

Think about how you could use a multimedia presentation to present what you have learned from the article and other research. Plan your multimedia presentation in the space below.

EXPLAIN
THE PHENOMENON

How were humans involved in this photo?

Summarize It

Explain how humans affect Earth's water resources.

_____

_____

_____

_____

_____

_____

_____

_____

_____

REVISIT

PAGE KEELEY SCIENCE PROBES

Revisit the Page Keeley Science Probe on page 21. Has your thinking changed? If so, explain how it has changed.

Three-Dimensional Thinking

1. What will most likely happen if lake water becomes polluted by humans?

   A. Animals in the lake will die.

   B. There will be more fish in the lake.

   C. The pollution will not hurt the plants or animals in the water.

   D. It will change the soil around the lake into pebbles.

2. How can we use conservation to help preserve water resources? Circle all that apply.

   A. Take shorter showers

   B. Collect rainwater to water indoor plants

   C. Dump dirty water into the sewers

   D. Turn off the faucet while brushing my teeth

   E. Shower at the same time every day

Extend It

As you are walking home from school one day, you notice someone throw an empty plastic bottle. It bounces into the nearby creek. While you can't chase after the person or the water bottle, what can you do to make sure something like this doesn't happen more often in the future? Write or draw what you could do below.

_____

_____

_____

_____

_____

KEEP PLANNING

STEM Module Project
Engineering Challenge

Now that you have learned about human impact on water resources, go to your Module Project to consider this information as you design a rainwater collection system.

Beach Changes

Three friends go to the same beach together every summer. They notice one year that the dunes are much smaller than before, and that the walk from the street to the shoreline is much shorter. They wonder what is happening to the beach. This is what they said:

Cecelia: *I think the dunes and beach just look smaller because we are older.*

Omar: *I think the dunes and beach are smaller because too many people are walking on it.*

Nari: *I think the dunes and beach have been worn away by wind and water.*

Who do you think has the best idea? _____

Explain your thinking.

_____

_____

_____

_____

You will revisit the Page Keeley Science Probe later in the lesson.

Effects of the Hydrosphere

ENCOUNTER
THE PHENOMENON

What changes the shape of the land?

🜚 GO ONLINE

Check out *Natural Bridges* to see the phenomenon in action.

Look at the photo and watch the video of the natural bridges. What questions do you have about the phenomenon? Record or illustrate your thoughts below.

Did You Know?

Most natural bridges or sea arches start out as a cave.

INQUIRY ACTIVITY

Wave Erosion

Think about how the natural bridges were formed. Water on Earth affects other parts of Earth's surface.

Make a Prediction How does the energy of beach waves affect Earth's materials, including soil, sand, gravel, and small rocks? Record your predictions in the center column in the data table on the next page.

Carry Out an Investigation

BE CAREFUL Wear safety goggles to protect your eyes. Use caution to avoid spills.

1. Gather your materials. Place the soil, sand, gravel, and small rocks on one end of each of the separate aluminum pans.

2. Pour 250 mL water into the other end of each pan.

3. Place a textbook behind one of the pans. Lift the pan slowly only to the height of the textbook and set it back down on your desk to cause waves. Observe the effect the waves have on the Earth material.

4. **Record Data** Repeat step 3 three times with the same pan. Record your observations in the data table on the next page.

5. Repeat step 3 and step 4 with the other three pans.

💬 Talk About It

What other variables could you test besides the type of land material? Talk about it with a partner.

Materials

 safety goggles

 4 small aluminum pans

 250-mL measuring cup and water

 soil

 sand

 gravel

 small rocks

 textbook

Type of Land Material	How Will Water Move the Material?	Observations
soil		
sand		
gravel		
small rocks		

Communicate Information

6. Did your results support your prediction? Explain.

_____

_____

_____

 Describe how this **modeled** the **cause and effect relationship** between the water and the land materials.

_____

_____

Look for these
words as you read:

deposition

erosion

floodplain

glacier

Erosion and Deposition

The natural bridges were formed by erosion. **Erosion** is the process of weathered rock moving from one place to another. The process of eroded materials being dropped off in another place is **deposition**. Erosion and deposition work together to change the shape of the land.

Erosion and Deposition by Running Water

As water runs downhill, it can wash away soil and erode rock. The water, soil, and rocks will eventually flow into a larger body of water, such as a river. Rivers with fast-moving water tend to follow straight paths. Fast-moving water has more energy. It can wash away larger amounts of heavier sediment. Rivers with slow-moving water tend to follow curved paths. Slow-moving water has less energy. It carries smaller particles of sediment.

GO ONLINE Use the *Water Cycle* and *Weathering, Erosion, and Deposition* personal tutors to learn more about the effects of the hydrosphere.

The looping curves in this river are called meanders. Slow-moving water deposits sediment on the inside of a meander. Faster-moving water erodes sediment on the outside of meanders.

Rivers eventually flow into larger bodies of water, such as a lake or oceans. Since the water is no longer flowing downhill, it slows down. The sediment carried by the water is deposited on the bottom of the lake. Over time, this sediment builds up into a landform called a delta.

Rivers also deposit sediment when they flow out of a steep, narrow canyon. Here, the stream becomes wider and shallower. The water slows down as it spreads out. Sediment is deposited in a landform called an alluvial fan.

💬 Talk About It

Compare the images of the delta and alluvial fan with a partner.

⊗ GO ONLINE Watch the video *Earth's Hydrosphere* to see ways the hydrosphere interacts with Earth's other systems.

When water that is carrying sediment enters a larger body of water, the sediment is dropped, forming a delta.

When a rushing river runs out of a narrow canyon, it slows down and becomes shallower. Sediment is dropped, causing an alluvial fan to form.

INQUIRY ACTIVITY

Hands On

River Erosion

You have learned that strength and speed of moving
water can affect the shape of a river.

State the Claim How does the slope of a river affect how it
erodes materials? How does this affect the way a river looks?

_____

_____

_____

_____

Carry Out an Investigation

BE CAREFUL Wear safety goggles to protect your eyes.
Use caution to avoid spills.

1. Gather your materials. Divide the dirt evenly between the aluminum
   pans. Spread it evenly across the bottom of the pan.

2. Place one end of the first aluminum pan on one book, creating a slope.
   Place one end of the second aluminum pan on a stack of three books.

3. Pour a steady stream of water at the top of the elevated side of
   the first pan. Record your observations on a separate piece of paper.

4. Pour a steady stream of water at the same speed at the top of
   the second pan. Record your observations on paper.

5. **Record Data** Compare the paths left by each of the rivers you created.
   Explain what caused the difference in the shape of the river.

_____

_____

_____

_____

Floods

Water runs over the ground in streams and rivers. Sometimes, water enters a river faster than the river can carry it away. When water collects on land that is normally dry, it is called a flood. Floods occur when a body of water overflows banks or beaches. A flood may also occur during a heavy rainfall. Natural wetlands can soak up water and reduce the chances of a flood. Draining wetlands or cutting down plants along a riverbank may make floods more likely.

Floodwaters carry and deposit sediments over the land. A **floodplain** is a place that floods easily when river water rises. Floods can cause damage by carrying mud into homes and streets. However, floods can also have a positive effect on natural systems. After a flood, new soil deposits on the land. The nutrients in this soil help plants grow.

Hurricanes and Storm Surges

A hurricane is a very large, swirling storm that forms on the surface of tropical oceans. Strong winds, walls of clouds, and pounding rains are associated with these storms. When a hurricane moves toward a coast, winds and waves can force large amounts of water onshore. This event is called a storm surge. Flooding associated with storm surges and heavy rains can be severe.

A hurricane caused this flood along the Gulf Coast.

REVISIT Revisit the Page Keeley Science Probe on page 37.

Inspect

Read the passage *Erosion and Deposition by Glaciers*. Underline evidence that tells how glaciers form.

Find Evidence

Reread Where can you find most of the material that results from erosion and deposition by a glacier?

Highlight the text evidence that supports your answer.

Notes

_____

_____

_____

_____

_____

_____

_____

_____

_____

_____

_____

_____

_____

moraine

moraine

Erosion and Deposition by Glaciers

In very cold places, thick sheets of ice called **glaciers** creep over the land. Glaciers form where snow collects quickly and melts slowly. The weight on top of the mound puts pressure on the snow below. The snow in the lower layers of the glacier slowly turns to ice. Near the ground, the pressure from the snow above causes some of the ice to melt.

As the weight of the ice increases, the glacier begins to flow. The bottom and sides of the glacier freeze onto rocks. As the glacier moves, it tears rock from the ground. It scratches, flattens, breaks, or carries away things in its path. A glacier can make a valley wider and steeper. It can carve grooves into Earth's solid rock surface.

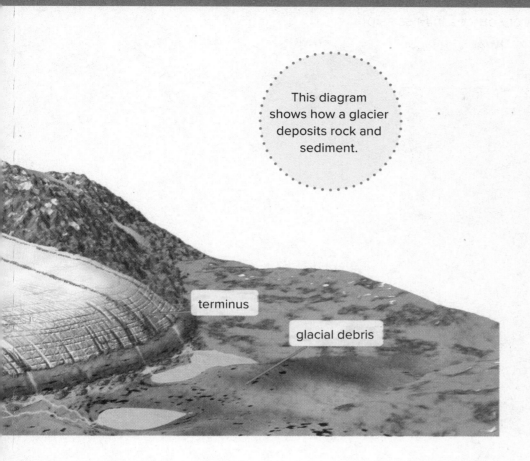

This diagram shows how a glacier deposits rock and sediment.

terminus

glacial debris

What information could a scientist learn from examining the materials left behind by a glacier?

Notes

As glaciers melt, they leave behind the rocks that were carried in the ice. The leftover rocks are called glacial debris. Glacial debris can be made of large boulders or small particles. It can have bits of gravel, sand, and clay.

The glacier drops most of its debris at the downhill end, or terminus. Long ridge-like mounds of rock called moraines can also be left behind by glaciers. Moraines are found at the terminus or along the edges of a glacier.

Study the diagram of a glacier. Describe how the glacier has caused changes to Earth's surface.

Copyright © McGraw-Hill Education

Mississippi Water Sources

Look at the map of different water sources in Mississippi.

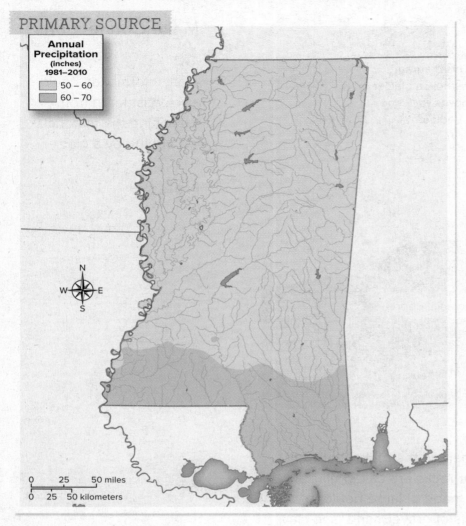

Annual Precipitation (inches) 1981–2010
- 50 – 60
- 60 – 70

0   25   50 miles
0   25   50 kilometers

1. What types of water sources do you see on the map?

_____

_____

2. Is there a pattern of water sources and average annual precipitation?

_____

_____

_____

A Day in the Life of a Water Engineer

From an interview with Monica Morales

Water engineers are civil engineers who focus on bringing water resources from lakes or groundwater to water treatment facilities, and then to homes and buildings. They might design the pipelines and pump stations that the water will follow. Water engineers use scientific studies to know if water is contaminated and to know how much water is needed by an area.

Water engineers get to be in the field when their pipeline designs are constructed. Some of the pipes in their designs are large enough for a car to drive through!

Water engineers are passionate about conservation efforts and educating others about how to care for water resources.

💬 Talk About It

What challenges might a water engineer face when working with farmers in different parts of the United States?

EXPLAIN
THE PHENOMENON | What changes the shape of the land?

Summarize It

Explain how the hydrosphere interacts with Earth's other systems.

_____

_____

_____

_____

_____

_____

_____

_____

_____

_____

_____

REVISIT
PAGE KEELEY SCIENCE PROBES
Revisit the Page Keeley Science Probe on page 37. Has your thinking changed? If so, explain how it has changed.

Three-Dimensional Thinking

1. How is erosion an effect of the hydrosphere? Circle all that apply.

   A. Erosion can be caused by moving water.

   B. Erosion can be caused by precipitation.

   C. The movement of glaciers causes erosion.

   D. The hydrosphere contains all of the land on Earth.

2. How does erosion shape the land?

   A. Earth's surface is changed by living things.

   B. Erosion does not change the shape of the land.

   C. Erosion happens only in the winter.

   D. Erosion carries the sediment and rock to another location, which changes the shape of the land.

3. Oceans are one of Earth's systems and can affect Earth's other systems.

   A. True

   B. False

Explain.

_____

_____

_____

_____

_____

_____

_____

_____

Extend It

You read a newspaper headline that says: *Record Number of Beachgoers Cause Beach to Wash Away.* Using what you know about the effects of the hydrosphere on erosion and deposition, write a letter to the editor explaining why people aren't the only reason a beach changes.

_____

_____

_____

_____

_____

_____

_____

_____

_____

_____

OPEN INQUIRY

How could you investigate the effect of the number of people on a beach and how they change the shoreline? Plan an investigation on a separate sheet of paper. Carry out your investigation if you can.

KEEP PLANNING

STEM Module Project
Engineering Challenge

Now that you have learned about the effects of the hydrosphere on Earth's other systems, go to your Module Project to consider this information as you design a rainwater collection system.

Design a Rainwater Collection System

How can rainwater be collected to be reused? You have been hired as a hydraulic engineer to design a rainwater collection system in your area. Use what you have learned to plan, build, and test your model. Use your results to communicate the advantages and disadvantages of using a rainwater collection system.

Planning after Lesson 1

Apply what you have learned about how water is distributed on Earth to your project planning.

What types of reservoirs are in your area? Research how much average rainfall your area gets in a year. Record the information below.

Planning after Lesson 2

Apply what you have learned about human impact on water resources to your project planning.

What do you need to consider about the effects of your rainwater collection system on Earth's systems?

Planning after Lesson 3

Apply what you have learned about how the hydrosphere affects Earth's other systems to your project planning.

Record information about where your rainwater collection system needs to be placed to not affect the area in a negative way.

Sketch Your Model

Use what you have learned to plan your rainwater collection system. Sketch your model in the space below. Identify the location of where you will build your model. Label the materials you will use to build your model and how water will flow through your design.

Design a Rainwater Collection System

Look back at the planning you did after each lesson.
Use that information to complete your final module project.

The Engineering Design Process

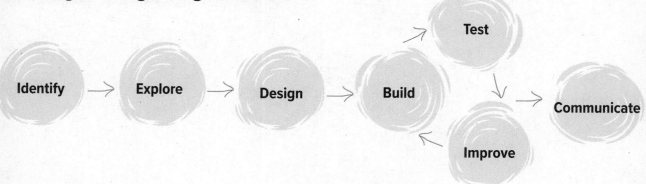

Identify → Explore → Design → Build → Test

Build → Improve → Communicate

Improve

Build Your Model

1. Identify the materials you will use to build your model.

2. Identify the location of where you will build your model.

3. Use the sketch of your model to build the model using the materials you identified in the location you have chosen.

4. Collect data from your model. Use the space on the next page to record your data.

Materials

_____

_____

_____

_____

_____

_____

_____

Test Your Model

MATH ❭ Connection Draw a data table in the space below to collect data about how well your rainwater collection system is working. Record data in the table over time.

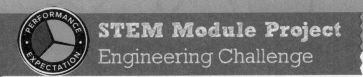
Communicate Your Results

Think about the advantages and disadvantages of using a rainwater collection system like the one you designed. Use evidence from your project to report on whether you think this is an effective solution.

Advantages of Rainwater Collection System	Disadvantages of Rainwater Collection System

Congratulations! You have completed your project. Think about how you could improve your model.

MODULE WRAP-UP

REVISIT
THE PHENOMENON

Using what you learned in this module, explain the effects of water on Earth, and how it can be collected to be reused.

Have your ideas changed? Explain.

Earth's Other Systems

ENCOUNTER
THE PHENOMENON

How are these flowers able to survive in such a harsh environment?

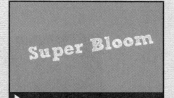
Super Bloom

◉ GO ONLINE
Check out *Super Bloom* to see the phenomenon in action.

💬 Talk About It

Look at the photo and watch the video of the "super bloom." What questions do you have about the phenomenon? Talk about them with a partner.

Did You Know?

A "super bloom" is a very rare occurrence in a desert environment.

Design a Desert Oasis

How can you plant a garden in the desert?
You have been hired as a landscape architect to
design a garden oasis in a desert climate.
Landscape architects plan, design, and take care of
natural environments for people to enjoy. They design
parks, gardens, and other important parts of
communities. At the end of this module, you will use
what you have learned to design a plan for your oasis.

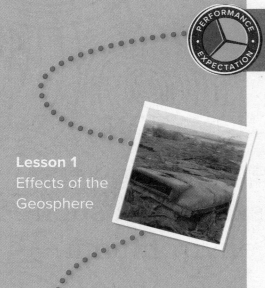

Lesson 1
Effects of the
Geosphere

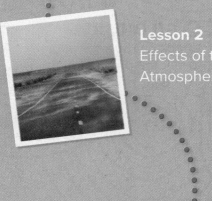

Lesson 2
Effects of the
Atmosphere

Lesson 3
Effects of the
Biosphere

Landscape architects must select materials that
will interact and grow successfully in the
environment where they are placed.

STEM Module Project

**Plan and Complete the Science
Challenge** Use what you learn
throughout the module to design and
communicate your plan.

The Solid Earth

Three friends were talking about the solid parts of Earth. They each had different ideas about where Earth's rock, soil, and sediments are found. This is what they said:

Francis: *I think the solid materials on Earth are found on land.*

Portia: *I think the solid materials on Earth are found under the oceans.*

Trent: *I think the solid materials on Earth are found on land and under the oceans.*

Who do you agree with most? _____

Explain why you agree.

_____

_____

_____

_____

You will revisit the Page Keeley Science Probe later in the lesson.

Effects of the Geosphere

ENCOUNTER

THE PHENOMENON

What happened to this abandoned school bus?

GO ONLINE

Check out *Flowing Lava* to see the phenomenon in action.

Talk About It

Look at the photo and watch the video of the abandoned school bus. What questions do you have about the phenomenon? Talk about them with a partner. Record or illustrate your thoughts below.

Did You Know?

Objects that become covered by the land around them can sometimes form imprint fossils.

INQUIRY ACTIVITY

Hands On

Predict the Flow of Water

Think about the bus covered in hardened lava. How can you model the effect of the shape of the land on the flow of liquid?

Make a Prediction What will happen when water is sprayed on a model of landforms?

_____

_____

_____

Materials

plastic tray or container

modeling clay

spray bottle

water

Carry Out an Investigation

BE CAREFUL Use caution to avoid spills.

1. Using the clay, build a model of different landforms on the bottom of your plastic tray.

2. Spray water evenly onto your model to simulate rain over a land surface.

3. Observe what happens to the water as it contacts the land surface.

4. **Record Data** Record or draw your observations in the space below.

Communicate Information

5. What happened to the land and the water in this model?

_____

_____

_____

_____

_____

_____

6. How did the land affect where the water went?

_____

_____

_____

_____

_____

_____

7. Did your observations support your prediction? Explain.

_____

_____

_____

_____

_____

INQUIRY ACTIVITY

8. How did this model demonstrate an interaction between the hydrosphere and geosphere?

_____

_____

_____

_____

_____

_____

_____

_____

How do **models** help scientists **obtain information** to understand the **interactions between Earth's systems**?

_____

_____

_____

_____

_____

_____

_____

GO ONLINE Explore *Water on Earth* to see more examples of how the geosphere and hydrosphere interact.

MAKE YOUR CLAIM

How does the geosphere affect the hydrosphere?

Make your claim. Use your investigation.

CLAIM

The geosphere affects _____.

Cite evidence from the activity.

EVIDENCE

The investigation showed that _____.

Discuss your reasoning as a class. Tell about your discussion.

REASONING

The evidence supports the claim because _____.

You will revisit your claim to add more evidence later in this lesson.

Features of the Geosphere

Look for these
words as you read:

hot spot

landslide

minerals

molten rock

The geosphere includes the solid and melted rock inside Earth. It also includes the soil, rock pieces, and landforms on Earth's surface. Scientists divide the Earth into several different layers. These layers include the crust, which is the rocky outermost layer. The mantle is just below the surface and made partly of **molten rock**. Molten rock is very hot melted rock. It can flow slowly like thick putty. The deepest layer of Earth at the center is the core. It is very hot, solid rock made mostly of iron. The inner core is solid.

Recall that a landform is a physical feature found on Earth's crust. These landforms change over time. Plate movement, weathering, and erosion can all change the appearance of Earth's surface.

Island Building

The Hawaiian Islands in the Pacific Ocean rest on a slowly moving plate. As it moves, the plate passes over a **hot spot**. A hot spot is an area where molten rock from deep within the mantle breaks through to the Earth's crust. Over millions of years, lava erupting from the underwater hot spot formed a mountain. Eventually, the mountain grew taller than the ocean's surface and became a volcanic island. As the plate moved, the island moved away from the hot spot, and a new island formed.

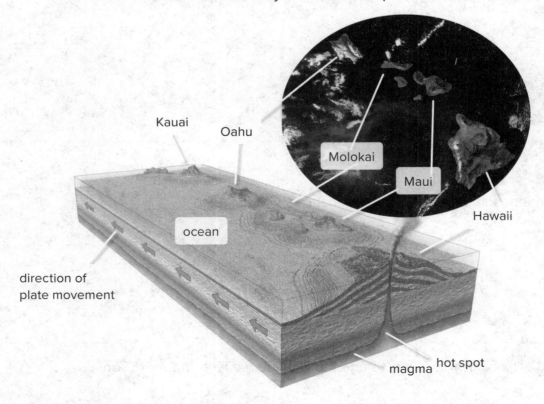

Kauai

Oahu

Molokai

Maui

Hawaii

ocean

direction of
plate movement

magma hot spot

Mountains

GO ONLINE Watch the video *Mountain Types* to help you understand the different types of mountains.

You have learned that huge plates of solid rock lie beneath the continents and ocean floor. These plates can move, which results in mountain building, earthquakes, and volcanoes.

Tension is a force that pulls things apart. It moves Earth's plates. Plates can also be moved by pushing forces. When plates are pushed together, the crust is forced upward, producing folded mountains.

The Himalayas in the Eastern Hemisphere are folded mountains. They began forming millions of years ago as India and Asia collided. As the plates continue to push into each other, the Himalayas grow about 5 millimeters (0.2 in.) taller every year.

When one plate rubs past another plate, the movement causes a force that twists, tears, or pushes one part of the crust past another. It can also cause Earth's surface to break apart along a continental boundary. When this happens, one side of the boundary moves up and the other side moves down. This movement produces fault-block mountains. The Sierra Nevada mountain range contains examples of fault-block mountains. When one plate is pushed below the other, a line of volcanic mountains can also form, such as the Cascade Range.

folded mountains

fault-block mountains

Volcanoes

Volcanoes form on land and on the ocean floor. You have learned that a volcano is an opening in Earth's crust. Volcanoes are located only at certain places on Earth's surface. Most volcanoes are found on the ocean floor.

However, volcanoes do not erupt at all continental boundaries. After collecting data about the directions in which parts of Earth slowly moved, scientists concluded that volcanoes tend to erupt where one plate is pushed under another plate. The plate melts under extreme heat and pressure as it is pushed down into the mantle. The melting forms magma, which pools in a chamber underneath the crust.

The magma may rest quietly for hundreds or thousands of years. Sometimes a crack forms above the lava chamber, or the pressure in the chamber becomes too strong to be held in by the rock above it. Then the magma rushes up toward Earth's surface.

An active volcano is one that is currently erupting or has recently erupted. A volcano that has not erupted for some time, but that scientists think may erupt in the future, is called a dormant volcano. A volcano that scientists think will not erupt again is an extinct volcano.

▶ **GO ONLINE** Explore *The Parts of a Volcano* to see how a volcano works.

The rocks on this cliff are unstable. Signs like this one warn people of the danger.

Landslides

When rocks and soil on a slope are loosened, gravity pulls them downward. A **landslide** is a sudden movement of a large amount of rocks and soil down a slope. Sometimes earthquake vibrations cause a landslide. Volcanic eruptions and heavy rains can also cause landslides. Human activity, such as clearing the land of trees in hilly areas, can make landslides more likely to occur.

When it rains, some water soaks into the ground. At some point, the ground can no longer absorb water. Then the remaining water mixes with the soil and forms mud. Eventually, the mud holds so much water that it becomes very heavy and cannot stay on the slope. If a lot of mud flows down the slope, it can knock down trees and destroy whatever is in its path. This event is called a mudslide, and it can cause rapid erosion of rocks and soil.

Earth's Soil

There are dozens of different kinds of soils found on the Earth's crust. Soil is made up of sand, silt, clay, nonliving plant and animal material, and **minerals.** Minerals are solid, nonliving substances found in nature. Water, air, and living things are also found in soil.

Different types of soil have different properties. Soils can be different colors. They can also have particles of different sizes. Various types of soil behave differently when you add water. Sandy soils hold very little water while medium-textured soils soak up water and hold air very well. Remember, many plants need to grow in soil. The different properties of soil result in different groups of plants being able to grow.

How Soil Forms

Soil can take thousands of years to form. As rocks break down, they become sediment. Plants take root and break down more rock. Animals move and mix the sediment. When plants and animals die, bacteria and fungi decompose them, adding humus to the soil. Humus is nonliving plant or animal matter. Humus adds nutrients to the soil for plants to grow. Living things allow the soil to renew year after year.

 Talk About It

Look at the map on page 75. What types of soil are in Pennsylvania? How might this affect the types of plants that grow in the state?

 COLLECT EVIDENCE

Add evidence to your claim on page 69 about how the geosphere interacts with the hydrosphere.

Desert plants are adapted to grow in sandy soils.

Medium-textured soils are good for growing many crops.

Some kinds of grapes grow well in clay soils.

Types of Soil

Each type of soil supports different plant and animal life. Most of the United States is covered by one of three types of soil: forest soil, desert soil, or grassland and prairie soil.

Forest Soil The soil in a forest has a thin layer of topsoil with little humus. Topsoil is home to many living things. Frequent rainfall carries minerals deep into the ground. Plants need long roots to reach these minerals. Much of the forest soil in the United States is in the Northeast and Southeast regions.

Desert Soil Desert soil is sandy and does not have much humus. However, desert soil is rich in minerals. Little rain falls to wash the minerals away. Animals can sometimes be raised in areas with desert soil. Crops can be grown only if water for the plants is piped to the area. Desert soil is found in the Southwestern region.

Grassland and Prairie Soil Grasslands and prairies are found between the Rocky Mountains and eastern forests. Crops, such as corn, wheat, and rye, grow on land from Texas to North Dakota. The soil is rich in humus, which provides nutrients for crops, and holds water so minerals are not washed deep into the ground.

Soil Types in the United States

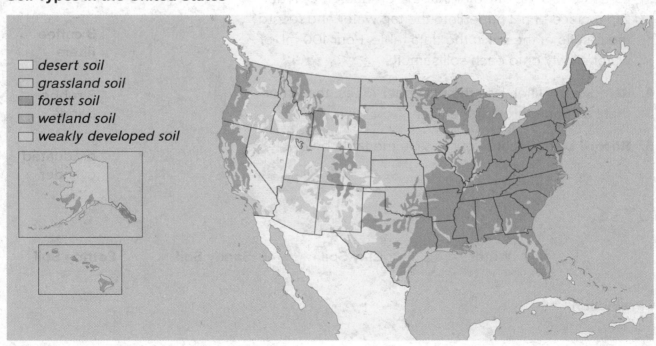

desert soil
grassland soil
forest soil
wetland soil
weakly developed soil

REVISIT Revist the Page Keeley Science Probe on page 63.

INQUIRY ACTIVITY

Hands-On

Effects of Soil on Water pH

Soil is part of the geosphere. Different types of soil have different properties. Investigate how the pH of soil affects the pH of water.

Make a Prediction How will three soils—clay, sandy, and potting soil—affect the pH of water? Record your predictions in the chart below.

Carry Out an Investigation

BE CAREFUL Wear safety goggles. Use caution to avoid spills.

1. Secure a coffee filter onto each jar using the rubber bands. Make sure there is enough filter in the jar to hold the soil.

2. Scoop two tablespoons of one type of soil into a coffee filter. Repeat for the other two types of soil.

3. Measure 100 mL of water into the graduated cylinder. Dip a piece of pH paper into the tap water and record the results of the pH in the data table. Pour 100 mL of water slowly onto each soil sample.

4. Remove the filter. Use the pH paper to test the water that has gone through the filter.

5. **Record Data** Record your data in the table below.

Materials

 safety goggles

 3 plastic jars

 pH paper

 metal tablespoon

 clay soil

 sandy soil

 potting soil

 3 coffee filters

 3 rubber bands

 graduated cylinder

	Water	Clay Soil	Sandy Soil	Potting Soil
Prediction				
pH				

6. Use the data that you collected during the activity to make a bar graph to communicate the effects the soil had on the pH of water. Make sure you label your axes with the pH level and the type of soil.

Communicate Information

7. Analyze your data. Did any of the soils affect the pH of the water? Explain your findings.

_____

_____

_____

💬 Talk About It

Did your findings support your prediction? Talk about it with a partner.

What Does a Cave Tour Guide Do?

Cave tour guides present historical and natural information about caves to guests. They provide information about the natural history of caves and how the caves form over many years. Cave tour guides must be able to communicate well with others and have to be aware of safety risks in caves. Cave tour guides may need to use special gear to stay safe when exploring caves. They may also have to be able to operate a boat to navigate tours through underground rivers.

Most cave tour guides must have at least a high school diploma. Experience working in a park setting is helpful to become a guide. Cave tour guides should also have knowledge of the natural history of the area in which they work. Once you become a cave tour guide, you may have to take classes on safety and navigation of caves.

It's Your Turn

What do you wonder about how caves form? Ask a question and perform research to find the answer to your question.

![FOLDABLES®]

Cut out the Notebook Foldables given to you by your teacher.
Glue the anchor tabs as shown below. Use what you have learned
to define the terms.

Glue anchor tab here.

Glue anchor tab here

▶ **GO ONLINE** Watch the video *Earth's Geosphere* to observe more examples of how the geosphere interacts with Earth's other systems.

Review

EXPLAIN
THE PHENOMENON

What happened to this abandoned school bus?

Summarize It

Explain how the geosphere interacts with other systems on Earth.

_____

_____

_____

_____

_____

_____

_____

_____

_____

_____

REVISIT
PAGE KEELEY
SCIENCE PROBES

Revisit the Page Keeley Science Probe on page 63. Has your thinking changed? If so, explain how it has changed.

Three-Dimensional Thinking

1. Explain one of the interactions of the geosphere and another one of Earth's systems that you learned about in the lesson. Include how this interaction results in change over time.

_____

_____

_____

_____

_____

_____

_____

_____

_____

2. Which is an example of a process in the geosphere that causes slow changes?

   A. earthquakes

   B. glaciers

   C. volcanoes

   D. landslides

Extend It

The geosphere includes the ocean floor. Over ninety-five percent of the ocean floor has not been explored by humans. Ask a question about the ocean floor that you would like to know the answer to. Use multimedia to organize your research and present it to the class.

_____

_____

Make a plan to research the answer to your question.
Record your research below.

_____

_____

_____

_____

_____

_____

_____

_____

_____

_____

KEEP PLANNING
STEM Module Project
Science Challenge

Now that you have learned about the effects of the geosphere, go to your Module Project to explain how the information will affect your design of a desert oasis.

Where Do Thunderclouds Come From?

Three friends noticed thunderclouds forming. It looked like a thunderstorm might be coming. They wondered where the thunderclouds came from. They each had a different idea:

Julie: *I think strong winds blew them here from another state.*

Doug: *I think they sank down from heavy clouds high up in the sky.*

Carlos: *I think they formed from warm air rising up into the sky.*

Who do you think has the best idea? _____

Explain your thinking.

_____

_____

_____

_____

You will revisit the Page Keeley Science Probe later in the lesson.

Effects of the Atmosphere

ENCOUNTER
THE PHENOMENON

What is causing this sandstorm to occur?

⊙ GO ONLINE

Check out *Sandstorm* to see the phenomenon in action.

Look at the photo and watch the video of the sandstorm. What questions do you have about the phenomenon? Record your thoughts below.

Did You Know?

'Dust devils' can form in sand and dust storms. A dust devil is like a mini tornado that forms and ends quickly!

INQUIRY ACTIVITY

Hands On

Warm and Cold Air Masses

Think about the sandstorm you just saw. The way the atmosphere moves can help predict the weather. Air is a fluid, just like water is a fluid. Using water of different temperatures, you will investigate how air masses move.

Make a Prediction What happens when air masses of different temperatures meet?

_____

_____

_____

_____

_____

Carry Out an Investigation

1. Wrap the piece of cardboard in aluminum foil. Place the cardboard in the center of the plastic bin so it forms a wall.

2. Add three drops of red food coloring into the measuring cup of warm water.

3. Add three drops of blue food coloring into the measuring cup of cold water.

4. Hold the cardboard tightly against the bottom of the bin as you pour the warm water into one side of the wall and the cold water into the other side of the wall.

5. **Record Data** Observe the water from the side as you remove the wall. Record or draw your observations in the box at the top of the next page.

6. Repeat the investigation with water of the same temperature, and only one side with food coloring. Record your observations.

Materials

 piece of cardboard

 aluminum foil

 clear plastic bin

 2 measuring cups

 4 cups of warm water

 4 cups of cold water

 red food coloring

 blue food coloring

Record Data

Communicate Information

7. Think about the water cycle. Could a cold air mass meeting a warm air mass cause precipitation? Explain.

_____

_____

_____

 How did this investigation **model** an **interaction** of the atmosphere and hydrosphere?

_____

_____

_____

💬 Talk About It

Did your results support your prediction? How does tracking air masses help predict the weather?

Earth's Atmosphere

Recall what you explored about air masses. The temperature of the atmosphere can determine activity within it. Even though air in Earth's atmosphere looks empty, it contains matter. The air particles in the atmosphere have mass and weight. There are different layers in Earth's atmosphere, which vary in temperature.

Weather is the condition of the atmosphere at a given place and time. Weather can vary depending on the time of day, season, or location. Weather can involve different forms of precipitation. When water vapor in clouds cools, it condenses and falls to the ground as rain, hail, sleet, or snow.

A shelf cloud like this one is a good sign that a strong line of storms will be moving through the area.

The air mass that is passing over an area affects the weather in that area. An **air mass** is a large region of air that has a similar temperature and humidity. Depending on where they form, air masses can be cool, warm, dry, or humid.

When one air mass meets a different air mass, the meeting place is called a front. A front is the boundary between two air masses that have different temperatures. Along fronts, weather can change rapidly. Look at the diagram below to see the difference between three different types of fronts: warm fronts, cold fronts, and stationary fronts.

Different Fronts

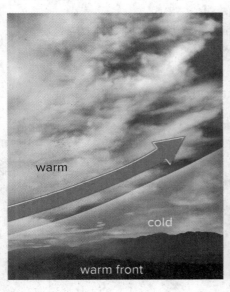

warm

cold

warm front

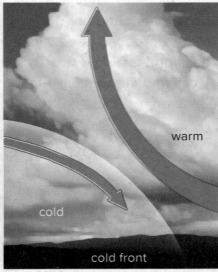

warm

cold

cold front

cold

warm

stationary front

Read a Diagram The arrows in the diagram indicate temperature as well as the direction of movement. Red arrows indicate warm air movements. Blue arrows indicate cold air movements.

Warm Fronts A warm front forms when a warm air mass pushes into a cold air mass. The warm air goes up and over the cold air mass. It often brings light, steady rain.

Cold Fronts A cold front forms when a cold air mass pushes under a warm air mass, forcing the warm air quickly upward. It often brings stormy weather.

Stationary Fronts Sometimes rainy weather lasts for days. This is caused by a stationary front, which is a boundary between air masses that does not move.

Weather Events

Most weather events are storms. A storm is a violent disturbance in the atmosphere. Storms involve sudden changes in air pressure, which cause rapid air movements.

GO ONLINE Explore *How a Thunderstorm Forms* to learn more on the stages of this weather event.

Thunderstorms are rainstorms that include lightning and thunder. They are the most common type of severe storm. The dangers in thunderstorms are lightning, strong winds, and flash floods.

Tornadoes and Derechos are the strongest types of thunderstorms. A tornado is a rotating, funnel-shaped cloud with wind speeds of up to 512 kilometers per hour (318 miles per hour). They can change direction abruptly, moving in one direction and then another. Tornadoes can cause terrible damage, breaking buildings, uprooting trees, and lifting cars into the air. A derecho is a widespread, long-lasting windstorm that occurs with some thunderstorms. They cause damage similar to tornadoes, but the damage occurs in one direction on a straight path.

REVISIT Revisit the Page Keeley Science Probe on page 83.

Tropical Storms occur near the equator where the ocean is warm. A tropical storm is considered a hurricane when winds get higher than 119 kilometers per hour (74 miles per hour). Hurricanes are dangerous storms, causing coastal flooding and severe wind damage. From space, a hurricane looks like a spiral of clouds with a hole in the center, called the "eye." The fastest winds and heaviest rains occur next to the eye.

Winter Storms occur when a cold, dry air mass meets a warm, humid air mass. Snowstorms such as blizzards happen when snow or sleet occur with high winds and cold air temperatures. Ice storms occur when rain falls and the ground temperature is cold enough that ice forms on outside surfaces. Winter storms can cause power outages, so it is important to be prepared with supplies before a winter storm occurs.

GO ONLINE Watch the video *Earth's Atmosphere* to learn more about the atmosphere's effects.

Some areas in the United States can experience heavy snowfall. The region around Lake Tahoe, California can see snow as early as September and as late as May.

Copyright © McGraw-Hill Education Stas Volik/Shutterstock

Inspect

Read the passage *Climate*. Underline text evidence that explains the factors that determine an area's climate.

Find Evidence

Reread the passage. What is the difference between weather and climate? Highlight the text evidence that supports your answer.

Notes

_____

_____

_____

_____

_____

_____

_____

_____

_____

_____

_____

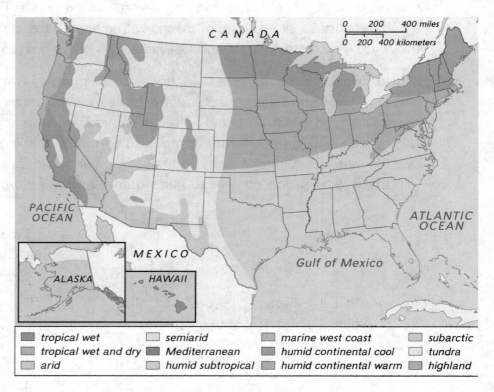

■ tropical wet	□ semiarid	■ marine west coast	■ subarctic
■ tropical wet and dry	■ Mediterranean	■ humid continental cool	□ tundra
□ arid	■ humid subtropical	■ humid continental warm	■ highland

Climate

Weather changes every day. However, the weather in any area tends to follow a pattern. **Climate** is the average weather pattern of a region over time. Climate varies from place to place. Important variables in determining climate are average temperature and average rainfall.

Climate is related to latitude, which is a location's distance north or south of the equator. Because of Earth's shape, areas closer to the equator receive more intense energy from the Sun than areas farther from the equator.

Climate is also related to how the hydrosphere and atmosphere interact. The distance a location is from a large body of water affects its temperature and average rainfall throughout the year. The geosphere also affects climate. Areas with mountains at higher altitudes have cooler climates.

garytog/iStock/Getty Images

One way to describe an area's climate is by the plants that grow there. Plant types require different levels of precipitation, sunlight, and temperature. For example, temperate forests have different weather during the four seasons. Temperate trees, such as oak trees, respond by losing leaves before the cold, dry winters.

ENVIRONMENTAL ⟩ Connection

How does climate affect plants?

_____

_____

_____

_____

_____

_____

Make Connections
💬 Talk About It
What are some climate patterns you can observe in the United States? Use the map to talk about it with a partner.

Notes

_____

_____

_____

_____

_____

_____

_____

_____

_____

_____

_____

_____

_____

Chance of Extreme Weather

Think about where you live. What types of weather events have you experienced in your area?

Make a Prediction How likely are different weather events to happen in the area where you live?

_____

_____

_____

Carry Out an Investigation

1. Describe the typical weather in your area during each season throughout the year in the table below.

Weather in Each Season			
Winter	**Spring**	**Summer**	**Autumn**

2. Research historical data of weather events in your area. Record the data in the table below.

Type of Weather Event			
Thunderstorm	**Tornado**	**Winter Storm**	**Tropical Storm**

Communicate Information

3. Use the data to summarize the weather events that are most likely to occur in your area. Why do you think this is the case?

_____

_____

_____

_____

4. Use the data to summarize the weather events that are least likely to occur in your area. Why do you think this is the case?

_____

_____

_____

How Do You Become an Imagery Analyst?

Imagery Analysts study images to gather information. One field they may specialize in is the weather. They use technology to collect images and then share their findings with others. Photographs, maps, and videos taken by weather satellites, like the one pictured above, are just a few examples of images they study. Imagery Analysts share their findings with other agencies such as the government and television news stations.

Imagery Analysts need at least a one-year degree from a technical school or community college. By attending school, students can use technology to learn how to study different types of imagery. They learn how to use technology such as radar and satellite images. After studying images, students learn to professionally share their findings both verbally and through written reports.

It's Your Turn

What do you wonder about how weather satellites work? Research information online or at your library to find out more about this powerful technology.

Data Analysis

Idaho Elevations

Idaho has a wide variety of different types of landforms.
The landforms found in Idaho can affect the weather in each region.

Make a Prediction How is the weather affected by elevation?

Carry Out an Investigation

1. Look at the two maps side by side.

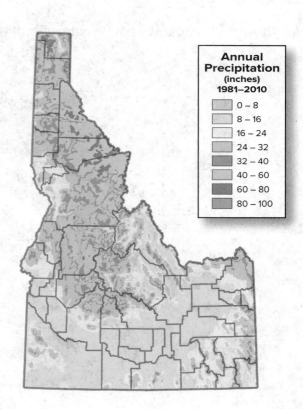

 Talk About It

Analyze Data Is there a relationship between the elevation of
landforms and precipitation in an area? Explain why or why not.

Review

EXPLAIN
THE PHENOMENON

| What is causing this sandstorm to occur?

Summarize It

Explain how the atmosphere interacts with other systems on Earth.

_____

_____

_____

_____

_____

_____

_____

_____

_____

REVISIT
PAGE KEELEY SCIENCE PROBES

Revisit the Page Keeley Science Probe on page 83. Has your thinking changed? If so, explain how it has changed.

Three-Dimensional Thinking

City	Average Temperature in January	Average Snowfall in January
Albany, NY	-5°C (23°F)	45.7 cm (18 inches)
Tahoe City, CA	-2°C (29°F)	101.6 cm (40 inches)
Reno, NV	2°C (36°F)	15.2 cm (6 inches)

1. What can you interpret about the data in the chart?

    A. Albany, NY is cold in January with a lot of snow.

    B. Tahoe City, CA is cold in January with little snow.

    C. Reno, NV is cool in January with a lot of snow.

2. How is evaporation a cause of precipitation?

Extend It

Think about the effects that the atmosphere has on Earth's other systems. How could you develop a plan to be prepared for an extreme weather event? Outline the steps of your plan below. Draw a diagram if you need to.

_____

_____

_____

_____

_____

_____

_____

_____

_____

KEEP PLANNING

STEM Module Project
Science Challenge

Now that you have learned about the effects of the atmosphere, go to your Module Project to explain how the information will affect your design of a desert oasis.

LESSON 3 LAUNCH

Use of Resources

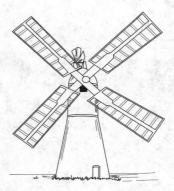

Four friends were talking about how humans use renewable and nonrenewable resources. They each had different ideas. This is what they said:

Wyatt: *When humans use renewable resources, they can harm the environment. Nonrenewable resources do not harm the environment.*

Sarah: *When humans use renewable resources, they do not harm the environment. Nonrenewable resources can harm the environment.*

Hector: *Both renewable and nonrenewable resources can harm the environment when humans use them. Just because some resources are renewable doesn't mean they have no harmful effects.*

Gus: *Renewable and nonrenewable resources do not harm the environment if they are natural resources. Natural resources are not harmful to the environment.*

Who do you agree with most? _____

Explain why you agree.

_____

_____

_____

_____

You will revisit the Page Keeley Science Probe later in the lesson.

Effects of the Biosphere

ENCOUNTER
THE PHENOMENON

How might human activity affect wildfires?

GO ONLINE

Check out *Wildfire* to see the phenomenon in action.

Talk About It

Look at the photo and watch the video of the wildfire. What questions do you have about the phenomenon? Talk about them with a partner. Record or illustrate your thoughts below.

Did You Know?

Four out of every five wildfires in the United States are started by humans.

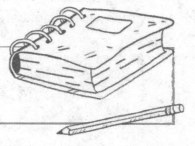

INQUIRY ACTIVITY

Overfishing

GO ONLINE

Think about the wildfire you saw. Other human activities affect ecosystems. Explore the *Overfishing* simulation to learn about the balance in a lake ecosystem.

Make a Prediction How will the number of fish caught relate to the number of fishing boats used over time?

_____

_____

_____

_____

Carry Out an Investigation

1. For each of the first five fishing seasons, choose only one boat to send out to fish in the lake. How many total fish were caught?

   _____

   _____

2. For each of the next five fishing seasons, choose two boats to send out to fish in the lake. How many total fish were caught?

   _____

   _____

3. For each of the next five fishing seasons, choose three boats to send out to fish in the lake. How many total fish were caught?

   _____

   _____

Communicate Information

4. How did the fishing boats affect the population of fish in the lake ecosystem?

_____

_____

_____

5. What number of boats resulted in the greatest number of fish caught? Do you think this could last?

_____

_____

_____

_____

6. What other factors might affect the fish population? Consider factors that did not appear in this model of a lake ecosystem.

_____

_____

_____

_____

 Talk About It

Did the results support your prediction? Talk about it with a partner.

> ▶ **GO ONLINE** Watch the video _Earth's Biosphere_ to see examples of interactions of Earth's systems.

deforestation

endangered

extinct

Earth's Biosphere

Remember that Earth's biosphere consists of all of the living things on Earth. You have learned a lot about how plants and animals interact with Earth's systems. Humans are also part of the biosphere and affect Earth's air, water, land, and other living things.

Recall that living things use the many resources that Earth provides. A natural resource is a material found in nature that humans and other living things can use. These resources are either renewable resources or nonrenewable resources. Renewable resources, such as water, wind, and sunlight, can be replaced in nature or will not run out. Nonrenewable resources are natural resources that are found in a limited amount on Earth's surface and will eventually run out.

Even though some human activities can harm natural resources, we can make choices to protect them. You have learned that conservation is a great way to help protect the natural resources on Earth. There are many other simple ways to help protect natural resources, including the following ideas:

- Organize a group to pick up trash. Dispose of all trash and recyclables properly.

- Plant new trees, bushes, and flowers that are native to the area.

- Compost garbage, grass, and leaves. Use the compost to feed plants instead of using chemical fertilizer.

- Ride a bike, walk, or take public transportation instead of riding in a car.

ENVIRONMENTAL ⟩ Connection

How do you make decisions to protect natural resources?
List two examples.

_____

_____

REVISIT Revisit the Page Keeley Science Probe on page 101.

PAGE KEELEY
SCIENCE
PROBES

Protecting Plants and Wildlife

🔵 **GO ONLINE** Explore *Living Things in Ecosystems* to learn more about how living things interact with Earth's land, water, and air.

Plants and animals get what they need to survive from Earth's resources. Still, there are many that do not get what they need to survive and are considered **endangered**. This means they are at risk of becoming **extinct**, or dying out completely. Human activities are partly to blame for this, such as pollution in the form of litter and acid rain. These can especially affect ocean ecosystems, which are home to a wide variety of organisms.

Another threat to ecosystems because of human choices is deforestation. **Deforestation** is the removal of large areas of trees. These trees are used to make goods, and the cleared land is used to build structures. The animals that live in the forests are forced to move. They may not be able to find another habitat that has the resources they need.

Remember that invasive species can have a serious impact on the environment. Humans can make smart choices when they plant new trees that are species native to the area. This way, the trees can provide balance to all of Earth's systems in that habitat without bringing disease or affecting the soil.

Endangered species such as this condor are tracked by scientists.

The effects of deforestation are difficult to undo, since it takes new trees a long time to grow and mature.

INQUIRY ACTIVITY

Endangered and Extinct

Changes in ecosystems can change the populations of the animals that live there.

Make a Prediction How can changes on Earth affect the population of animals over time?

_____

_____

_____

Carry Out an Investigation

1. Which animals from the table are extinct? How do you know?

_____

_____

_____

2. Which animal population is growing? Use online and print resources to find more information about how this animal population is being protected and is able to grow.

_____

Type of Animal	Population *All populations are estimates.*		
	1970	**1990**	**Present**
Mountain gorilla	250	600	880
Leatherback sea turtle	90,000	70,000	55,000
Golden toad	unknown	1	0
Whitefin dolphin	400	100	0

 Talk About It

What do you think is the cause of the changes in these animal populations? Discuss your ideas with a partner.

Interactions Everywhere

Think about everything you have learned about Earth's systems and how they interact. In the table below, cite examples for how the different parts of the biosphere affect Earth's other systems.

Effects of the Biosphere on Earth's Systems			
Parts of the Biosphere	Atmosphere	Geosphere	Hydrosphere
Plants			
Animals			
Humans			

GO ONLINE Explore the simulation *Biosphere of a Tide Pool* to see the interactions of Earth's systems within a tide pool ecosystem.

Protecting Against Forest Fires

Humans have a responsibility to help protect living things in ecosystems and to limit the amount of resources used. Scientists continue to research ways to reduce damage to ecosystems.

State the Claim What can we do to protect against forest fires?

_____

_____

_____

Carry Out an Investigation

1. Read the Investigator article *Mediterranean Cypress Trees*.

2. Use what you read and information from other sources to develop a plan to reduce the spread of a forest fire. Consider the materials, cost of materials, and time it will take to implement your plan.

3. Write your plan on a separate piece of paper to send to a national park. Provide a diagram if it helps you communicate your plan.

💬 Talk About It

Share your plan with a partner. Compare and contrast your plans and use what you discuss to make improvements to your plan.

Use the **information you have obtained** about the role of humans in the biosphere and **the effect of the biosphere on other Earth systems.** Explain how **developing and using a model** can help humans make better decisions about the **effect** they have on the environment.

_____

_____

_____

_____

_____

Curator of Birds and Mammals

Interview with Dudley Wigdahl

How did you become interested in this career?

I grew up at the beach. I had a natural attraction to studying science. Some important teachers recognized this, and their positive reinforcement helped push me in this direction. I loved being outside and studying the animal habitats around me. I enjoyed animals and caring for pets.

How would you describe your career? What does a typical day of work look like?

I have spent a fascinating 42 years working directly with all kinds of marine mammals and birds. No two days are ever the same and I use my education in my job every day. Currently, I spend a lot of time in my office working with staff to oversee the care for our animals. When I first started out, I spent more time directly caring for the animals and making sure they stayed healthy.

How do you use science and engineering in your career?

I learned about animals during school. I use what I was taught to educate young visitors at the aquarium. I use math all the time when figuring volumes, diameters, radius, and water chemistry. I use engineering skills to design animal enclosures.

It's Your Turn

Would you consider becoming a curator of birds and mammals? Talk about it with a partner.

Review

EXPLAIN
THE PHENOMENON

How might human activity affect wildfires?

Summarize It

Explain how the biosphere, including humans, affects Earth's hydrosphere, geosphere, and atmosphere.

_____

_____

_____

_____

_____

_____

_____

_____

REVISIT Revisit the Page Keeley Science Probe on page 101. Has your thinking changed? If so, explain how it has changed.

Three-Dimensional Thinking

1. A farmer discovers a large amount of a toxic material on the edge of her property, near an empty highway. She worries that the pollution will run downhill to the nearby lake if it starts to rain. Which will most likely happen if lake water becomes polluted by the toxic material?

   A. Animals in the lake will die.

   B. There will be more fish in the lake.

   C. The pollution will not hurt the plants or animals in the water.

   D. The pollution will change the soil around the lake.

2. Models are useful tools used by scientists and engineers to observe cause and effect. Why is it important for scientists and engineers to use models to study the effects of the biosphere on Earth's hydrosphere, atmosphere, and geosphere? Circle all that apply.

   A. By using a model, scientists and engineers can pretend that the biosphere does not affect Earth's other systems.

   B. Models can help scientists and engineers observe the effects of the biosphere on other systems over time, but in a short amount of time.

   C. By using a model, scientists and engineers can observe potential harmful effects without actually causing harm to living things or Earth's other systems.

   D. Models are easier to read than written reports.

Extend It

Think about an effect of the biosphere on Earth's other systems that is a result of human behavior. You are interested in getting involved with your local government to help change the behavior. Write a speech that you could share at city hall or with the town council to persuade the politicians to write a bill to help make the behavior happen less often. Use multimedia to help you present your speech to the class, who can pretend to be the town council.

_____

_____

_____

_____

_____

_____

_____

_____

_____

_____

_____

_____

_____

KEEP PLANNING

STEM Module Project
Science Challenge

Now that you have learned about the effects of the biosphere, go to your Module Project to explain how the information will affect your design of a desert oasis.

Design a Desert Oasis

How can you plant a garden in the desert?
You have been hired as a landscape architect
to design a garden oasis in a desert climate.
Use what you have learned in each lesson
to design a plan for your oasis.

Planning after Lesson 1

Apply what you have learned about the effects of the geosphere on
other systems to your project planning.

What part of the desert would be part of the geosphere?
How does this affect the other systems within the desert?

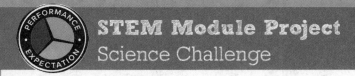
Planning after Lesson 2

Apply what you have learned about the effects of the atmosphere on other systems to your project planning.

How does the atmosphere affect other systems in the desert? What do you need to consider to plan your desert oasis?

Planning after Lesson 3

Apply what you have learned about the effects of the biosphere on other systems to your project planning.

What parts of a desert are part of the biosphere? How can parts of the biosphere help you plan your desert oasis?

Design a Desert Oasis

Look back at the planning you did after each lesson. Use that information to complete your final module project.

Design your plan for a desert oasis in the space below. Label the parts of the oasis and what part of Earth's systems they belong to. Use your model to describe how the different systems interact with one another.

Communicate Your Results

Using your design plan, define a design problem that a city in a desert region might need to solve in order for people to live there. Remember that the solution to the problem cannot be too expensive for the city to be successful.

_____

_____

_____

_____

_____

_____

_____

_____

_____

_____

_____

_____

_____

_____

_____

_____

_____

_____

_____

_____

_____

MODULE WRAP-UP

REVISIT
THE PHENOMENON

Using what you learned in this module, explain how the super bloom is able to occur.

How has your thinking changed? Explain.

Science Glossary

A

abiotic factor a nonliving part of an ecosystem

acid rain harmful rain caused by the burning of fossil fuels

air mass a large region of air that has a similar temperature and humidity

algae bloom a sometimes harmful rapid increase in the amount of algae found in water

apparent motion when a star or other object in the sky seems to move even though it is Earth that is moving

atmosphere the gases that surround Earth

B

bacteria a type of single cell organism

biosphere the part of Earth in which living things exist and interact

biotic factor a living thing in an ecosystem, such as a plant, an animal, or a bacterium

C

chemical change a change that produces new matter with different properties from the original matter

chemical property a characteristic that can only be observed when the type of matter changes

climate the average weather pattern of a region over time

colloid a type of mixture in which the particles of one material are scattered through another without settling out

condensation the process through which a gas changes into a liquid

conductivity ability for energy, such as electricity and heat, to move through a material

conservation the act of saving, protecting, or using resources wisely

conservation of mass a physical law that states that matter is neither created nor destroyed during a physical or chemical change

constellation any of the patterns of stars that can be seen in the night sky from here on Earth

consumer an organism that cannot make its own food

D

decomposer an organism that breaks down dead plant and animal material

deforestation the removal of trees from a large area

deposition the dropping off of eroded soil and bits of rock

E

endangered when a species is in danger of becoming extinct

energy the ability to do work or change something

energy flow the movement of energy from one organism to another in a food chain or food web

erosion the process of weathered rock moving from one place to another

evaporation a process through which a liquid changes into a gas

extinct when a species has died out completely

F

floodplain land near a body of water that is likely to flood

food chain the path that energy and nutrients follow among living things in an ecosystem

food web the overlapping food chains in an ecosystem

fungi plant-like organisms that get energy from other organisms which may be living or dead

G

galaxy a collection of billions of stars, dust and gas that is held together by gravity

gas a state of matter that does not have its own shape or definite volume

geosphere the layers of solid and molten rock, dirt, and soil on Earth

glacier a large sheet of ice that moves slowly across the land

gravity the force of attraction between any two objects due to their mass

groundwater water stored in the cracks and spaces between particles of soil and underground rocks

H

habitat a place where plants and animals live

hot spot an area where molten rock from within the mantle rises close to Earth's surface

hydrosphere Earth's water, whether found on land or in oceans, including the freshwater found underground and in glaciers, lakes, and rivers

I

ice caps a covering of ice over a large area such as in the polar regions

invasive species an organism that is introduced to a new ecosystem and causes harm

L

landslide the sudden movement of rocks and soil down a slope

light year the distance light travels in a year

liquids a state of matter that has a definite volume but no definite shape

M

magnetism the ability of a material to be attracted to a magnet without needing to be a magnet themselves

mass the amount of material in an object

matter anything that has mass and takes up space

meteor a chunk of rock from space that travels through Earth's atmosphere

meteorite A meteor that strikes Earth's surface

minerals solid, nonliving substances found in nature

mixture a physical combination of two or more substances that are blended together without forming new substances

molten rock very hot melted rock found in Earth's mantle

moon phases the apparent shapes of the Moon in the sky

N

nitrogen cycle the continuous circulation of nitrogen from air to soil to organisms and back to air or soil

O

orbit the path an object takes as it travels around another object

oxygen-carbon cycle the continuous exchange of carbon dioxide and oxygen among living things

P

phloem the tissue through which food from the leaves moves throughout the rest of a plant

physical change a change of matter in size, shape, or state that does not change the type of matter

physical property a characteristic of matter that can be observed and or measured

planet a large, round object in space that orbits a star

precipitation water that falls from clouds to the ground in the form of rain, sleet, hail, or snow

predator an animal that hunts other animals for food

prey animals that are eaten by other animals

producer an organism that uses energy from the Sun to make its own food

R

reflectivity the way light bounces off an object

reservoir an artificial lake built for storage of water

revolution one complete trip around an object in a circular or nearly circular path

rotation a complete spin on an axis

runoff excess water that flows over Earth's surface from a storm or flood

S

solid a state of matter that has a definite shape and volume

solubility the maximum amount of a substance that can be dissolved by another substance

solution a mixture of substances that are blended so completely that the mixture looks the same everywhere

star an object in space that produces its own energy, including heat and light

stomata pores in the bottom of leaves that open and close to let in air or give off water vapor

storage the process of water being stored on Earth's surface in the ground or as a water feature

T

tides the regular rise and fall of the water level along a shoreline

transpiration the release of water vapor through the stomata of a plant

V

volcano an opening in Earth's surface where melted rock or gases are forced out

volume a measure of how much space an object takes up

W

water cycle the continuous movement of water between Earth's surface and the air, changing from liquid into gas into liquid

weather the condition of the atmosphere at a given place and time

X

xylem the plant tissue through which water and minerals move up from the roots

Index

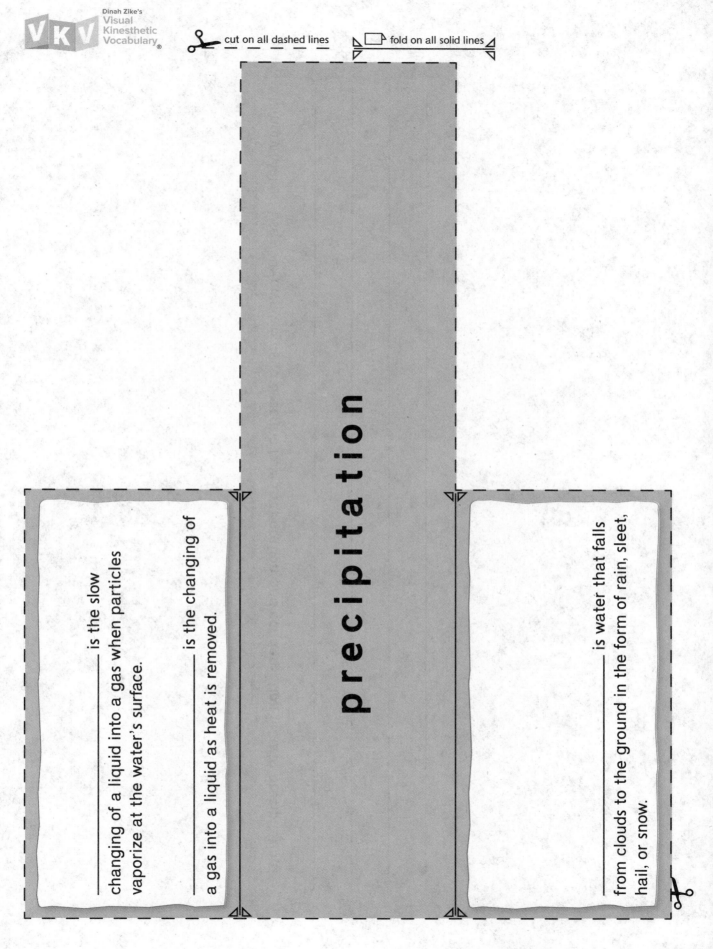

precipitation

_____ is the slow
changing of a liquid into a gas when particles
vaporize at the water's surface.

_____ is the changing of
a gas into a liquid as heat is removed.

_____ is water that falls
from clouds to the ground in the form of rain, sleet,
hail, or snow.

✂ cut on all dashed lines   ▭ fold on all solid lines

Memory Maker: Use your own words to describe how **evaporation** and **condensation** lead to **precipitation**.

condensa

evapora

✂ cut on all dashed lines   ▭ fold on all solid lines

hydrosphere

The _____ is the part of Earth in which living things exist and interact.

The _____ is the layers of rock, dirt, and soil on Earth, including the mantle, cores, and crust.

The _____ is Earth's water, whether found on land or in oceans, including the fresh water in ice, lakes, rivers, and underground.

Module: Earth's Other Systems   **VKV3**

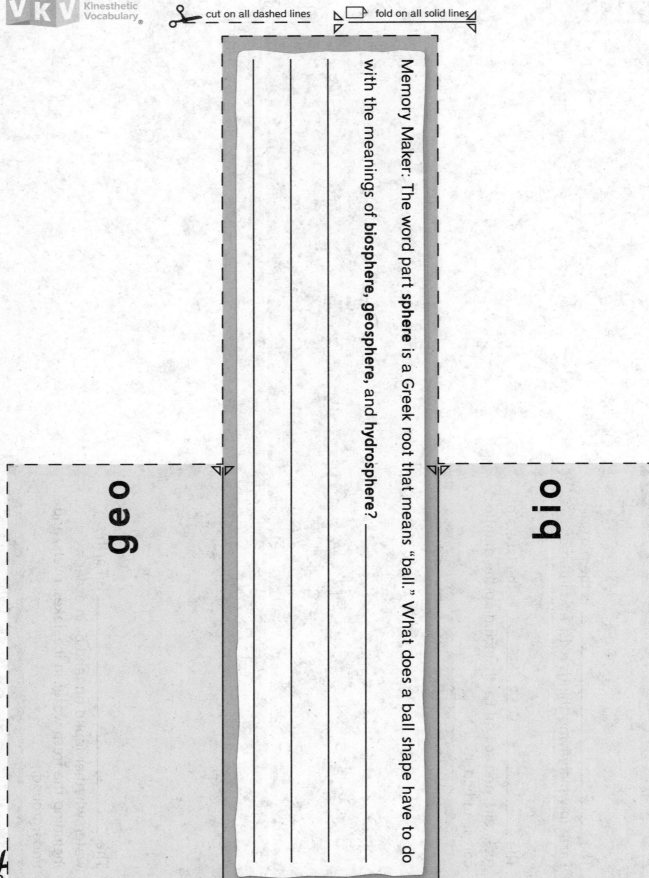

geo

bio

Memory Maker: The word part **sphere** is a Greek root that means "ball." What does a ball shape have to do with the meanings of **biosphere, geosphere,** and **hydrosphere?**